Colourful Birds of Australia

Colourful Birds of Australia

Text by Neil Hermes

Front cover: Rainbow lorikeets.
Back cover: Rufous night-heron. (See page 27.)

This second edition published in 1990 by PR Books
5 Skyline Place, Frenchs Forest, NSW, Australia, 2086
Telephone: (02) 975 1700, Facsimile: (02) 975 1711
First edition published in 1988 by
Child & Associates Publishing Pty Ltd
A wholly owned Australian publishing company
This book has been edited, designed and typeset in Australia
by the Publisher
Typesetting processed by Deblaere Typesetting Pty Ltd
Text by Neil Hermes
Captions by Dalys Newman
Photographs from the Australian Picture Library
© Australian Picture Library 1988, 1990
Printed in Hong Kong
Also published by Child & Associates Pty Ltd as part of
Australia: Land of Colour

National Library of Australia
Cataloguing-in-Publication data

Hermes, Neil.
    Colourful birds of Australia.

    2nd ed.
    Includes index.
    ISBN 1 875113 28 2.

    1. Birds—Australia. 2. Birds—Australia—Pictorial works.
    I. Title.

598.2994

Colourful Birds of Australia

To the casual observer, Australia's brilliantly coloured parrots, the flightless emu and perhaps the giant of the kingfishers, the kookaburra, characterise our unique birds. However, when we look beyond the showy colours, the large size and the raucous laughter of these familiar species, Australia's birdlife has even more of interest to offer.

Many groups of birds are found in Australia and nowhere else in the world. This is so of the lyrebirds, plains wanderer, treecreepers, magpies, currawongs, magpie-lark, bowerbirds and emu. Other groups, although found elsewhere, are well represented in the island continent and these include penguins, albatrosses, tropicbirds, parrots and cockatoos. Not only are the types of birds of great interest but their behaviour patterns set them apart from birds elsewhere in the world. Who could not be amazed by the mallee fowl which lays its eggs in a huge compost heap and then carefully tends the mound to ensure that the correct temperature is maintained, or by the bowerbird which constructs an elaborately decorated display arena to lure a mate.

These unique qualities of Australia's birdlife can be attributed in part to the long isolation of the continent from the rest of the world. In addition, Australia's changing climate over millions of years has created new opportunities that birds have exploited. For example, as the inland became drier some species adapted to a desert way of life and others became restricted to the wetter coastal zones.

Of Australia's 700 or so bird species, over half are found nowhere else in the world. Some are exceedingly rare and are the subject of rescue programmes, for example, the noisy scrub-bird in Western Australia and the Lord Howe Island woodhen. Other native species have benefited from European settlement of the landscape and become problems to pastoralists or orchardists, for example, the silvereye. Unfortunately, over twenty species of birds from other parts of the world have been introduced by man into Australia and many of these are serious pests both to agriculture and to native birds, for example, the common starling, common mynah and feral pigeon.

Australia's largest bird

The emu is a distinctive Australian and Australia's largest bird. It shares a position on the Australian Coat of Arms with the red kangaroo. Although flightless, emus are not handicapped by this. Not only can they walk long distances to find food (distances of many hundreds of kilometres), they can also run at speeds of up to 50 kilometres per hour. Pairs form in the summer months and the female lays up to twenty eggs during the winter. She then abandons the male who does all the caring for the eggs and chicks. During the eight weeks of continuous incubation he may lose up to 8 kilograms in weight. After the eggs hatch he then cares for the distinctly striped young for up to a year and a half. In Western Australia large numbers of emus move from the drier pastoral zone into wheat country, causing considerable damage. Many forms of control have been attempted by the farmers. However, the most dramatic, and the least successful, were the infamous 'Emu Wars' of 1932. An army squad with two Lewis guns set up a series of emu drives which killed less than 200 birds at a cost of 10 000 rounds.

Cassowaries, large flightless relatives of the emu, are found only in thick rainforests of northern Queensland and New Guinea. The Australian cassowary is a handsome bird. The feathers are black, head and neck are blue and a pair of red wattles hang from the throat. The head is topped with a large casque. An adult can stand up to 2 metres tall. Cassowaries are generally solitary birds preferring the safety of the thick forests. Occasionally, when the natural rainforest fruits are in short supply, they will enter gardens and eat cultivated fruits.

Cassowaries are known to be aggressive and can kill people. They are strong kickers and the foot is armed with a lethal spike. Although aggressive behaviour is usual only during the breeding season some individuals appear to be generally belligerent.

We now turn from the flightless cassowary of the tropical jungles to the southern ocean and the largest flying bird alive today—the wandering albatross. With a wingspan of some 3.5 metres and its effortless flight the wandering albatross must be the 'ultimate' flying machine. Albatrosses are typically birds of the southern oceans, although three species do occur in the northern Pacific Ocean and individual birds have been rarely recorded in the North Atlantic. Eight species of albatross are known in Australian waters and most nest on islands in the sub-Antarctic. The wandering albatross is typical. It breeds in islands such as Macquarie Island well to the south of Tasmania. Wandering albatrosses nest every two years since it takes almost twelve months to raise one chick. If an egg is lost early in the season the adults may attempt to nest again the following year. Outside the breeding season the adults and immature birds leave the sub-Antarctic waters and feed in waters around southern Australia, South Africa and Tierra del Fuego. In the winter months it is not unusual to see concentrations of feeding albatrosses around the southern coasts of Australia. The concentrations off Sydney have been studied for many years by ornithologists. The researchers chase albatrosses in small boats and capture them with nets before they can get airborne. By placing numbered bands on the bird's leg much valuable information on the habits and wanderings of the birds has been gathered. Banding studies suggest that wandering albatrosses may live up to thirty-five years.

Similar studies have been conducted on the shearwaters or muttonbirds which nest on islands in southern Australia, particularly Bass Strait. Called muttonbirds because the young are harvested and eaten, shearwaters are dark brown seabirds with long, slender wings like albatrosses. Shearwaters nest in burrows dug into the grasslands and breed in vast colonies. The short-tailed shearwater has a remarkable life cycle. All females lay their single egg between 20 November and 1 December and many individuals lay on the same day every year. After hatching the chick is abandoned during the day and fed only at night. The chick can grow to double the size of the adult birds. The adults then abandon the chicks and they starve for a period of about two weeks before they emerge from the burrow and take to the wing and the sea. What then occurs is even more remarkable. The whole shearwater population migrates via New Zealand and the central Pacific Ocean to the waters north of Japan. Some birds reach the Bering Sea. Having spent the southern winter in the northern Pacific Ocean the adults return to the same Bass Strait island breeding burrow to lay on the same date again! Large numbers of young birds perish on their first migration which is a round trip of about 30 000 kilometres.

The harvesting of the chicks by the Bass Strait Islanders is a traditional activity which is now strictly controlled. Known numbers of birds are taken as chicks from the burrows and the effect on the total shearwater population has been calculated to be negligible. The chicks are plucked and sold either fresh, frozen or salted for human consumption. Oil from the stomachs is used to produce suntan lotion, the fat is given as a food supplement to cattle and the down is the principal source of down for Australian-made sleeping bags.

An unusual pelican

The Australian pelican is a popular and friendly bird. It often develops a particular understanding of our pastimes and learns where to wait for handouts of fish. The Australian pelican has behaviour patterns which make it unlike the pelicans that occur on most other continents. Its breeding is adapted to the irregular pattern of rainfall which characterises much of the dry Australian continent. Breeding may occur at any time of the year and may be delayed for years until suitable conditions are present. Large colonies nest on islands in Lake Eyre and other inland lakes after wet years. Unfortunately, breeding is often too successful and as the inland dries up, large numbers of young pelicans find it difficult to obtain food. At these times pelicans can become quite bold and appear on ponds in city parks or even on swimming pools in their desperate efforts to find food.

Flocks of pelicans often feed together in a highly organised manner. They swim slowly together and drive shoals of small fish into shallow water and then simultaneously dip their bills to catch their food.

Close relatives of the pelican are the tropicbirds. Three species live in the tropical oceans and islands of the world and two, the red-tailed and the white-tailed tropicbirds, occur in Australian waters. Tropicbirds are very tame during nesting and will allow close approach. This, added to coloured bills and black and white plumage, makes them distinctive and well-known seabirds. In flight their tail streamers, which may be up to half a metre in length, add to their appeal.

Storks are another group of birds with great appeal even if they are given legendary capacity beyond the biologically possible! Typically, storks are large, long-legged black and white birds with brilliant red colourings. This is all true of Australia's only stork, the jabiru. Found in marshland from the Kimberleys through the tropics and down the eastern coast to around Sydney, the jabiru is usually solitary or seen in pairs. This is unlike storks overseas which often form sociable flocks. The jabiru is a versatile feeder and lives on a range of aquatic animals such as fish, frogs, crayfish and lizards. Typical of storks everywhere, jabirus build large and bulky nests placed high in a tree.

The dancing brolga

Similar in stature to the jabiru, Australia's most common crane, the brolga, is a much more finely built and statuesque bird. In addition the brolga has an elegant and inspiring courtship dance. Songs and poems have been written in recognition of these inspiring displays. One Aboriginal legend claims that the brolga was once a famous dancer but having declined the advances of an evil spirit was changed to a stately brolga. The brolga's dances appear to be passionate affairs. The pairs face one another and bow and bob as they move towards and away from one another. Occasionally one bird will trumpet wildly with its head held back. The most spectacular routine is when they leap into the air kicking their legs forward and then gracefully glide back to earth. Remarkably, these performances do not occur during the breeding season but they are presumed to form a part of continued reinforcement of the pair bond.

Brolgas share the inland waterways with about twenty species of ducks, geese and swans. These waterfowl display an amazing range of adaptations to the varying reliability of rainfall across the continent. Along the eastern and southern coasts where winter rains are reliable and in the tropical north where the summer Wet Season is the norm, waterfowl have regular breeding cycles. However, through most of the continent, the inland ducks are erratic breeders and have adapted to respond quickly to flooding whenever it occurs.

The magpie goose is characteristic of the waterfowl of the tropical north. A distinctive large black and white goose it feeds on vegetation in swamps and billabongs. When the Wet Season comes, large tracts of low-lying country are flooded, producing vast fields of swamp plants suitable for young geese. Breeding occurs in huge concentrations and the nest is a floating platform of vegetation. The colonies form as protection from dingoes which take many young geese. As the Dry Season approaches, the adults and young head back to the permanent waterways. This yearly cycle is typical of other northern waterfowl such as the tree-ducks, Burdekin duck and pygmy geese.

In the coastal areas of southern Australia, where winter rainfall is reliable, there is a regular pattern of winter and spring breeding of some waterfowl species, for example, the maned duck, Australian shelduck and chestnut teal. These species usually nest in hollow trees near permanent water.

A curious group of ducks, the diving ducks, are also regular breeders. The musk duck and blue-billed duck live on permanent deep waterways. They feed on small aquatic animals which they catch while diving. They can remain submerged for up to 60 seconds. Since they live on large permanent waterways breeding can occur on a regular annual basis.

The remaining Australian ducks, including the grey teal, black duck and rare freckled duck, breed only in response to favourable weather conditions. Concentrations of these birds occur on the large inland lakes and rivers and, as rains fall and flooding begins, so breeding commences. It can be at any time of the year. These species are also quite adaptable in their nest requirements. The grey teal will nest on the ground, in trees, under rocks and even in rabbit burrows! If flooding occurs the lack of a suitable nest site will not deter the grey teal from breeding.

Continued on page 41 . . .

The sulphur-crested cockatoo (Cacatua galerita). Flocks of thousands of these white cockatoos with fine sulphur-yellow crests are commonly seen out on the plains and in open forest country.

The colourful king parrot (Alisterus scapularis) grows to 45 centimetres and is fairly common in coastal forests and near rivers. This bird nests in large trees and timber milling is endangering its habitat.

A ground-feeding bird, the long-billed corella (Cacatua tenuirostris) is found in open timber country and is becoming increasingly rare.▼

The young male red-tailed black-cockatoo takes four years to acquire his magnificent adult plumage. Common in forests and open timber country, the bird advertises its presence with an extremely rowdy call.

Probably the most plentiful parrot in Australia, the budgerigar (Melopsittacus undulatus) *forms into enormous flocks, flying in tight precise formation. They are a common sight on the open plains.*

Frolicsome and friendly rainbow lorikeets.

Rainbow lorikeets (Trichoglossus haematodus) *feed in the tops of flowering native trees and are conspicuous for their noisy screeching and incessant chatter.*

13

The eclectus parrot (Eclectus roratus) *is found mainly in forests and jungles in the northern regions of Australia.* ▲

Cockatiels are also known as cockatoo parrots and quarrions. They are frequently seen by the roadside, often on telephone wires, and have an attractive whistling call.

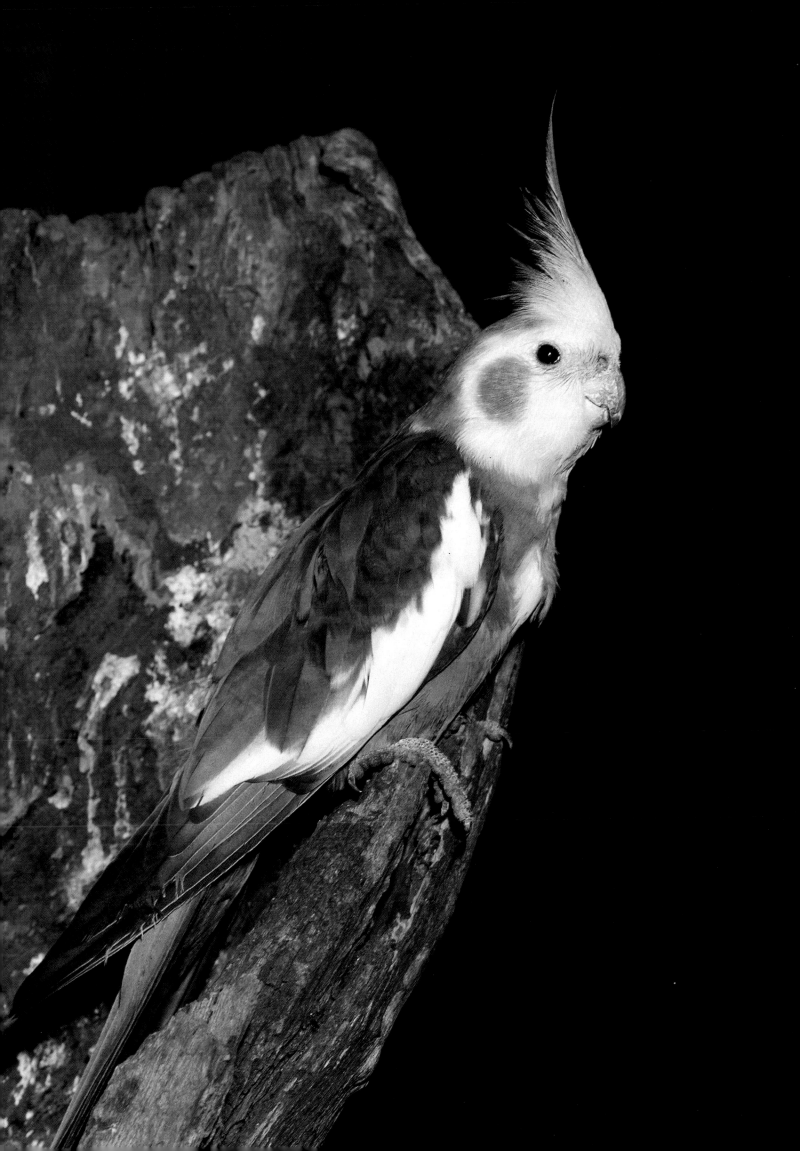

The pink cockatoo (Cacatua leadbeateri). *These birds are rare, found only in isolated spots in dry inland areas.*

Often seen at groves of tall banksias, the slow-flying yellow-tailed black-cockatoo tears pieces from rotting tree limbs to extract wood borers and grubs.

The largest of all the parrots, the palm cockatoo (Probosciger aterrimus) *has a powerful beak with which it cracks open nuts from palm trees. When disturbed the bird 'blushes', its cheek patches turning a deeper red.*

A flock of galahs takes to the sky. The entire flock will 'change colour' as they wheel, revealing the beautiful pink under-feathers. Today, the galah is one of the most prolific of inland birds.

The emu is among the few birds of which the female is liberated. She does the courting, the male looks after the young. The female lays a clutch of about nine eggs and the male sits on them for about eight weeks until the chicks emerge. He is then also responsible for their upbringing.

Emus (Dromaius novaehollandiae) are Australia's largest bird and the world's second largest after the ostrich. Almost always found in flocks, stalking majestically across inland plains or open forest country, emus are non-flyers but can run as fast as a galloping horse. (Previous page.)

The pied butcherbird (Cracticus nigrogularis) is so-called because of his habit of 'hanging up' his meat. Often he will tightly wedge his dead prey, or other items of food, into tree forks.

Bird of a thousand voices, the superb lyrebird (Menura novaehollandiae) is the prince of mockingbirds—it mimics all bush noises. The bird's outstanding feature is the tail of the mature male which is made up of sixteen striking feathers. The male rakes up a dancing mound about a metre wide from soft soil and struts around it to impress the female, a small homely bird lacking any adornment. (Following page.)

Kori bustard (Ardeotis kori). These birds wander the dry interior of the continent looking for places that have had heavy rain. They feed on seeds, fruits, small insects and even mice and lizards.

King of Australia's kingfishers, the laughing kookaburra (Dacelo novaeguineae) is also called the laughing jackass. Its laughter resounds at sunset and sunrise throughout the bush, in towns and many suburban areas.

Although belonging to the kingfisher family, the kookaburra rarely catches fish, preferring to dine on insects, snakes, rodents, lizards, yabbies and the young of other birds. ➤

The richly coloured sacred kingfisher is commonly found in mangrove and open timber country, near creeks and rivers. It announces its arrival with a ringing four-syllable call, a repeated 'ki'.

The noble wedge-tailed eagle (Aquila audax), *Australia's largest bird of prey.*

Increasingly found in urban areas, the white-plumed honeyeater (Lichenostomus penicillatus) *is a noisy, pugnacious fellow whose distinctive 'chick-o-wee' is a common garden call.*

The bold, inquisitive Lewin honeyeater (Meliphaga lewinii). All honeyeaters possess a brush-like tongue for sucking honey from blossoms.

The rufous night-heron patrols shallow swamps, lake shores and river edges during the night, hunting for fish, frogs and crustaceans which it catches with its long pointed bill.

Cattle egrets (Ardea ibis) *have introduced themselves to Australia from Asia since 1940. They often feed around grazing stock.*

The brolga, Australia's only native crane, has had a great influence on Aboriginal life and culture. The bird's dancing has inspired the choreography of several corroborees, and the dancers' dress and make-up imitate the big colourful bird. ▼

Australasian gannets—big birds of the sea. Talented divers, gannets will plunge into the sea from great heights of 20 metres or more.

Often seen soaring high above the coastline and inland rivers, the white-breasted sea-eagle (Haliaeetus leucogaster) *is a strong flyer with a wingspan of about 2.3 metres.*

A greedy bird, the great cormorant (Phalacrocorax carbo) *will sometimes eat until it is too heavy to take off.* ▼

The shy and elusive black-necked stork (jabiru) is the only member of the stork family found in Australia.

Sooty terns on Michaelmas Cay, Queensland. Often called sea-swallows, terns are smaller than gulls, with more finely shaped wings and beaks, and a typical swallow-like flight.

The beautiful sight of pelicans floating on calm waters has provided inspiration for many artists and poets. Pelicans are found throughout Australia. They feed in a close circle, constantly moving to stir the fish and prevent their escape at the same time. (J. Baker & APL.)

The black swan (Cygnus atratus) *is found throughout Australia except in the extreme north.
The bird's colour is accentuated by a red bill and white highlights on the wings.*

*Silver gulls—the scavengers of the air—are seen mainly on the seashores with
occasional appearances inland.*

Until the Dutch explorer Willem de Vlamingh discovered the black swan in Western Australia in 1697, the black-coloured swans were unknown. Today, in parks and gardens throughout the world, the Australian black swan is as well known as the white swan of Europe.

Looking like something left over from Halloween, the masked owl (Tyto novaehollandiae) is found in savannah country and is popular in some areas as a rabbit destroyer. ➤

The mallard duck (Anas platyrhynchos) has been introduced into Australia in small numbers and is found in parks and garden pools.

The tawny frogmouth (Podargus strigoides) *is an odd-looking bird with a wide gaping beak and soft plumage.*

The tawny frogmouth is one of nature's great camouflage artists. The birds can sit or crouch along the branch of a tree, blending perfectly into the overall picture.

Australia's most common owl, the southern boobook is a woodland bird whose pleasant 'book-book' call can be heard almost anywhere in the bush at night. (Following page.)

Australia's main game-shooting birds are the black duck and the grey teal. Since they are both species which have irregular breeding strict annual bag limits are set based on annual duck counts by wildlife authorities.

A strange swan

Perhaps one of Australia's most distinctive waterfowl and one which attracted the most interest from the first white settlers is the black swan. Since Northern Hemisphere swans are white, this black antipodean emphasised the stark differences of their adopted new land. The unfortunate black swan was under some criticism by the homesick Europeans. George Bass, the noted explorer, wrote:

That song, so celebrated by the poets of former times excellently resembled the creaking of an ale-house sign on a windy night.

Birds of prey, the hawks, falcons and eagles, have never been celebrated by the poets on their musical calls. In fact the calls of most raptors are at best whistles but usually harsh screams. Australia's largest bird of prey is the wedge-tailed eagle. This close relative of the North American golden eagle has a wingspan of up to 2.5 metres. Like its American cousin, the wedge-tailed eagle had a reputation for being a major predator of domestic livestock. Research has shown that although live lambs are occasionally taken they are usually sickly and dying anyway. Most lambs taken by eagles are usually already dead from other causes. Very occasionally a young eagle will take a healthy lamb but the impact on sheep production is quite insignificant. Wedge-tailed eagles prefer to feed on wallabies, rabbits or lizards. Carrion is an important component of the diet.

Australia boasts another large eagle, the white-breasted sea-eagle which floats over most coastal areas, patrolling for fish, sea-snakes, waterbirds and other small animals. This eagle shares the coastal skies with another fish eater, the osprey. This bird is very rare in other parts of the world and the subject of elaborate protection measures. Although not common in Australia, the osprey has not yet become an endangered species. The thick coastal rainforests are home to white goshawks and crested hawks. Further inland we find many species of kites and falcons.

Black kites are great scavengers and wherever you find a rubbish tip, abattoir or stockyard you will have a quota of these inquisitive birds. Inland bushfires often attract large flocks of black kites which prey on the scattering wildlife. The kites are often a hazard to fire spotters working from aircraft around the fire front.

The falcons are tenacious birds. The Australian hobby falcon, although only 30 centimetres in height, will attack much larger birds such as ducks and galahs. At the nest the fiery-spirited bird will attack any intruder. The author can vouch for its determination. To inspect a nest 15 metres up in a tree required protective clothing including a leather jacket, gloves, goggles and crash helmet!

The fascinating mallee fowl

A much safer bird to study is the remarkable mallee fowl. This large ground-dwelling inhabitant of the dry inland mallee scrub has fascinated scientists for decades. The mallee fowl's scientific name is Greek and literally means 'I abandon and desert my eggs'. On the surface it may appear that the mallee fowl is a vagabond but nothing could be further from the truth. The mallee fowl's life is constant toil. The female bird spends virtually all her time searching for food in order to produce her large eggs. This is exhausting work, for she may lay up to thirty-three eggs in the course of the summer breeding season and the country is dry and parched. The male has an even more demanding task, that is, to maintain the nest mound. The mound is a heap of rotting leaves, twigs and soil about 4 metres in diameter and up to 1.5 metres high. The eggs are laid into the centre of the mound. The male dutifully checks the temperature of the nest on a daily basis. In the spring when fermentation in the nest is high the mound is opened up to allow heat to escape. In summer, earth is heaped onto the mound to insulate the eggs. These excavations are all done by scratching and many cubic metres of material are moved back and forth each day. The tiny chicks never see their parents. After hatching, the chick digs unaided to the surface of the mound and immediately has to care for itself. It can fly within 24 hours! Life certainly wasn't meant to be easy for a mallee fowl!

One of the least known and rarest Australian birds is the plains wanderer. Although rather unspectacular in appearance, it looks a little like a quail and is of great scientific interest. The plains wanderer has no close relatives anywhere in the world and is classified in a group of its own. Its known home country is the grasslands of inland southern Australia, which are now heavily stocked with sheep, cattle and rabbits. The plains wanderer is probably vulnerable to attack by cats, dogs and foxes. This bird is now the subject of much research activity by the Royal Australian Ornithologists Union.

Another once rare bird, but one which has now been rescued, lives on the delightfully beautiful Lord Howe Island. The Lord Howe Island woodhen is a reddy-brown ground-dweller the size of a bantam. It is flightless since it had no natural predators on its island paradise. That is, until man arrived. The first settlers ate the birds and many sorry tales were recorded. In 1788 Thomas Gilbert wrote:

> Partridges [woodhen] ran in great plenty...several of these I knocked down, and their legs being broken, I placed them near to me as I sat under a tree. The pain they suffered caused them to make a doleful cry, which brought 5 or 6 dozen of the same kind to them, and by that means I was able to take nearly the whole of them.

Introduced cats and pigs soon removed most of the remaining birds from the island.

In 1979 only thirty woodhens remained and these were on the peak of the highest mountain. In 1980 nine birds were flown by helicopter from the mountain peak to a specially constructed compound on the island's lowland. Within three years ninety-two chicks were produced. As the chicks matured they were released to parts of the island from which pigs had been removed. Woodhens are now well established in many areas on the island where they hadn't been recorded in living memory. This great achievement is one of the few real success stories in attempting to rescue a very rare bird species through captive breeding.

'Help save the poor Bustard.' So reads the slogan on a promotional shirt produced by wildlife authorities trying to protect the little-known bustard. He is a large bird standing over a metre tall and with a wingspan of up to 2.5 metres. Bustards of one species or another occur throughout Africa, Europe, Asia and Australia. Being large ground-dwelling species that favour grasslands they are now threatened by many agricultural practices. In some parts of the world they are hunted for sport. Although bustards can fly well, they prefer to rely on camouflage for protection. When first disturbed by a predator or human, the large bird 'freezes' in the long grass. If the intruder makes a closer approach the bustard will cautiously stalk away still hoping to avoid detection. Only if really pressed does the bird finally start into a run and launch itself into flight. In Australia bustards are nomadic and most likely to be seen in the arid tropical north.

A small world

The twentieth century is supposed to be the century of travel. With the coming of large aircraft and relatively cheap fares, the world is now a smaller place. Australia is part of this world trend and is becoming a more popular destination for tourists. But Australia has been on the Asian tourist route for centuries. Every year, hundreds of thousands of international bird visitors fly into Australia from Asia and the Pacific. They are the migratory wading birds and range from the tiny red-necked stint weighing less than 25 grams to the eastern curlew measuring more than half a metre in length.

The migratory waders number over thirty different species and travel straight-line distances of up to 15 000 kilometres twice each year. A typical migratory wader is the sharp-tailed sandpiper. Weighing in at less than 100 grams it is perhaps more distinctly coloured than is typical of this notoriously plain group of birds. Its feet are yellow and its head rufous with the rest of its body boldly marked brown and black. Tens of thousands of these little birds feed on mudflats around the southern capitals during the Australian summer. As autumn approaches they feed furiously, adding about 30 per cent in weight in preparation for the long journey ahead. The sandpipers then form into large flocks and head north, probably via the Kimberleys or the Top End. From there we know little of their route through south-eastern Asia except that by May all the birds are on the breeding grounds in Siberia! Amateur ornithologists have captured and banded thousands of sandpipers. One bird caught near Sydney during February was found dead in April on the southern coast of mainland China! Amateur research efforts in Perth have resulted in returns from Siberia!

Other species of migratory waders travel annually between Japan and Australia. Since the long-term survival of these international bird species depends on human activities in countries in different hemispheres, international co-operation is required to protect them and their living places. The governments of Australia, Japan and China have agreements which commit themselves to these goals.

Terns are probably best known from the large colonies nesting on cays within the Great Barrier Reef. Some such as the sooty tern nest on the ground and others such as the white-capped noddy make seaweed nests in low trees. This bird is well known on some of the reef's resort islands. Terns feed on small fish which they catch by diving from a height above the water.

Australian parrots are among the most beautiful birds in the world. However, since many are very common, their beauty is often highly underrated. If the crimsom rosella, the rainbow lorikeet or even the galah was a rare bird its beauty would be highly praised. Such is the penalty of familiarity.

Australia has been called *Terra Psittacorum*—the Land of Parrots. Everyone has their favourite parrot, be it the bizarre palm cockatoo of the rainforests of Cape York or the confiding rosellas of a city garden.

The not-so-familiar budgie

Internationally the best known Australian parrot is the budgerigar. To one Aboriginal tribe it was the 'betcherrygah' to use one highly Anglicised spelling. To one of Australia's earliest and greatest ornithologists this small green parrot was known as the 'Warbling Grass Parakeet'. We can only wonder as to whether the budgie would ever have become a popular cagebird struggling under that name! Anyone used to only a single blue or white budgie living in a cage would be amazed to see the bird in its natural surroundings. Green buderigars are all that are known in the wild, no yellows or blues or whites. And where a single talkative bird is all that one would normally expect as a household pet, in the wild, flocks are numbered in the thousands. For the household bird a flight across the kitchen might be a marathon, but for its wild relatives hundreds of kilometres may be covered in the desert searching for seeding grasses and water. The contrasts of lifestyles could hardly be greater.

Sulphur-crested cockatoos are also popular cagebirds and this is due in part to their ability to mimic human speech. Having fed well, birds living well away from cities and towns often 'entertain' themselves by aimlessly tearing leaves and bark from trees. Unfortunately, this activity has been further developed by city-living cockatoos. They now 'entertain' themselves by tearing insulators from TV aerials, stripping putty from window seals or tearing up furniture or verandahs built of western red cedar. As can be understood, this has made sulphur-crested cockatoos extraordinarily unpopular birds with some people. They are viewed as being so destructive that they have been described as 'birds with bolt-cutters attached to their heads'. It is probably good practice in cities to observe sulphur-crested cockatoos from a distance and not encourage them to feed in suburban gardens. So long as the birds are busy looking for food less casual damage is likely to be caused.

Another popular parrot, which is often encouraged to feed at bird tables with no damaging results, is the rainbow lorikeet. Found along the eastern coast from Cape York to Adelaide this lorikeet has attained particular international stardom. At a reserve owned by the Queensland National Trust on Queensland's Gold Coast, thousands of rainbow lorikeets are fed every day. At eight in the morning and four in the afternoon the sky is filled with wheeling flocks of these brilliant blue, orange and green birds. They come in from the forests up to 30 kilometres away for their handouts of bread, honey and water. The birds are so keen to feed that it doesn't concern them who holds their food dishes. Every year thousands of people get pleasure, not only from seeing and photographing these brilliantly coloured birds at close range, but from having the birds perch all over them while they are feeding.

To small birds of the bush the melancholy call of the pallid cuckoo is a frightening sound. The calls continue during the season all day and all night. The cuckoo will not attack them but like cuckoos around the world it robs them of the chance to rear their own chicks. Over eighty different species of bush birds, including honeyeaters, whistlers and robins, play unwilling host to the cuckoo's offspring. The young cuckoo is so demanding that its adoptive parents will continue to push food into its gaping mouth despite its growing larger than themselves. Birds other than the foster parents are sometimes even allowed to give the bird a feed.

Australia's largest cuckoo is a very strange bird. The pheasant coucal was known to the early settlers as the swamp pheasant. It is a heavily built, long-tailed cuckoo which seems to crash from tree to tree rather than fly. Not only does the power of flight elude this cuckoo but it also rears its own young! Coucals are found in wet coastal country from Sydney north to beyond Darwin.

The cold nights are full of strange sounds besides cuckoos. Some of the owls have eerie calls described by their bush names; for example, the barn owl is also called the screech owl, and the barking owl is known to some bushmen as 'screaming woman'! Around the major cities the most frequently heard owl is the boobook or 'morepoke,' which has a repetitive two-part call like its name.

A distinctive large-eyed wide-mouthed night bird often mistaken for an owl is the tawny frogmouth. Although they can look alarming with their mouths open wide this bird is the master of camouflage. During the day a bird sitting motionless on a branch can look like a dead, broken tree branch.

The bushman's clock

Two species of kookaburra inhabit the Australian bush. The familiar laughing kookaburra or 'bushman's clock' is found along the coast from Cairns to Adelaide and Tasmania and Perth. The blue-winged kookaburra inhabits the tropical north, and as well as its bright blue wings it is distinguished by not laughing. The kookaburras are the largest kingfishers in the world. Most kingfishers, true to their name, live near water and eat fish. But not the kookaburras. They have adapted completely to a bush life and they live on insects, lizards and sometimes small snakes.

Kookaburras often become quite tame and line up in suburban gardens for handouts of meat or cheese. Once a regular beat is established parent kookaburras often bring their offspring along for a handout as well.

Another truly Australian bird which is well known for its song is the superb lyrebird. Although normally shy, in some places, notably Sherbrook Forest near Melbourne, the birds will allow visitors a close viewing. The male builds a display mound which is just a small clearing in the forest. When the winter breeding season approaches, the males use the display mounds to perform elaborate dances and song routines. The long tail is arched over the bird's head and produces a spray of feathers in the shape of a lyre.

The courting males prance about and vibrate the delicate tail feathers in a brilliant, shimmering display. A loud, rich song accompanies the dance. In addition to the male's own repertoire, he mimics any favourite sounds which he feels may impress his mate. It may include the songs of other birds or man-made sounds such as woodchopping or engine sounds. One remarkable bird learned to mimic a flute. Since the birds mimic one another the flute sound has now been incorporated into the repertoire of a group of neighbouring lyrebirds.

The focus of all this is to attract the female to the male's territory. Mating takes place on or near the mound and the female is then on her own to build the nest, incubate the eggs and raise the young.

Of all the brilliantly coloured small bush birds the ones which appeal to most people are the delicate fairy wrens. Around the capital cities of the south-east, the superb fairy wren makes his home. The breeding male is a cocky little bright blue and black. Like the female he has a long, narrow tail which is held at right angles to the body when perching. The females and immature males are a dull grey in colour and the female has a touch of red about the face. Since the superb fairy wren lives in family groups and only the mature male is brightly coloured it was often assumed that the one male had many females. This is not the case but it did not stop these wrens being called 'Mormon wrens'.

Wrens have a complex family structure which means they can produce many young when conditions are favourable. The one family group can produce many young in a year since the breeding male and the immature males and females all take part in caring for the young. Superb fairy wrens are friendly birds to have about a suburban garden but they are vulnerable to attacks from domestic cats.

Many of this continent's bushland birds share names with birds from the Northern Hemisphere. It must have made the early settlers feel less abandoned to give well-known bird names to the unfamiliar birds.

Hence Australia has robins, warblers, flycatchers, choughs and magpies which are quite unrelated to their northern namesakes. For some other new birds the apparent similarities were with several inhabitants of the Old World and we now have groups of Australian birds known as quail-thrushes and cuckoo-shrikes.

The Australian magpie family is a good example of how new names evolve. It contains a number of familiar and distinctive black and white birds—some are called magpies, after a superficial resemblance to magpies of the Northern Hemisphere; some are named currawongs, after their calls; and the butcherbirds are named for their eating habits.

The Australian magpie is both loved and hated. It is a large and inquisitive bird in the gardens and parks of many cities. They can become quite tame and enter houses for food. Unfortunately, magpies can become quite aggressive during the breeding season. Some birds defend the territory vigorously. The main method of defence is to swoop low over any intruder and let out loud calls of bill clacking. This can be quite off-putting and very annoying if the magpie's territory includes your backyard. Many curious techniques are observed to ward off swooping magpies. In Canberra, where this problem is acute, it is not uncommon to observe public servants walking to work holding an umbrella over their heads on a fine spring morning. Children similarly may be observed wearing hats, especially hats with 'eyes' drawn around the brim. The theory is that if magpies attack only from the rear, which seems usually to be the case, the novel headgear will deter the avian aggressor. Such are the birds (and people) of Australia.

Index

Numbers in *italics* denote colour photographs.

Acknowledgements

Henry Ausloos: page 36 above.
John Baker: page 33.
John Carnemolla: pages 12 below, 14, 16 above right, 17, 18–19, 20, 21, 24 above, 25, 29, 30, 32 above, 34, 36 below.
Colin Driscoll: pages 12 above, 16 above left.
Dallas and John Heaton: pages 9, 13, 22–3.
Leigh Hemmings: page 27 above.
Wade Hughes: page 28 above.
Gary Lewis: pages 26 below, 32 below, 35.
Jim McKay: page 16 below.
Fritz Prenzel: Front cover, pages 10, 11, 15, 24 below, 37, 38.
Derek Roff: pages 26 above, 27 below, 31, 39, 40.
Mike Warman: page 28 below.